1st Grade Jumbo Workbook
Spelling, Reading & Math

SPEEDY
PUBLISHING

Speedy Publishing LLC
40 E. Main St. #1156
Newark, DE 19711
www.speedypublishing.com

Spelling

E _ _ _ _ _ _ _

G _ _ _

F _ _ _

c _ _

A _ _ _ _ _

D _ _ _ _ _ _ _

$\mathcal{B}$ _ _

I _ _ _ _ _ _ _

J _ _ _ _ _

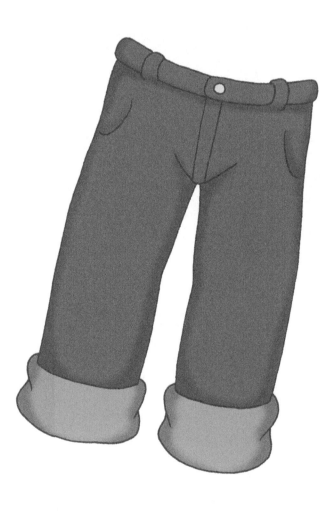

M _ _ _ _ _ _ _

$\mathcal{K}$ _ _

O _ _ _ _ _

L _ _ _ _

N _ _ _

T _ _ _ _ _

P _ _

S_ _ _ _ _

Q _ _ _ _

$\mathcal{R}$ _ _

W _ _ _

z _ _ _ _

*v*_ _ _ _ _

u _ _ _ _ _ _ _ _

y _ _ _ _

H_ _ _ _ _ _

Color and Identify

C _ t _ e _ p _ _ l a _

A _ l _ _ a _ o r

B _ _ _

E _ _ _ _ _ ant

F _ o _

G _ _ _

H _ _ _ e

J _ l l _ _ i s _

K _ n _ a _ _ o

L _ _ n

M o _ _ _

O _ t _ p _ s

R _ _ _ it

S _ _ _ e

Z _ _ _ a

Reading
and
Tracing

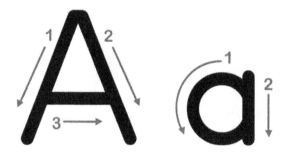

 Ant

 Arm

 Acorn

 Airplane

 Apple

B b

Bb Bb Bb

 Book

 Bee

 Ball

 Butterfly

 Backhoe

C c

 Car

 Cake

 Carrot

 Clock

 Cloud

D d

D ^{2} d

D d D d

 D uck

 D ragonfly

 D iamond

 D onut

 D olphin

E e

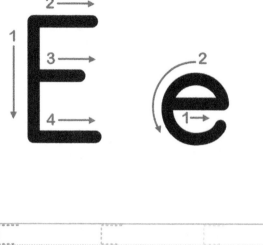

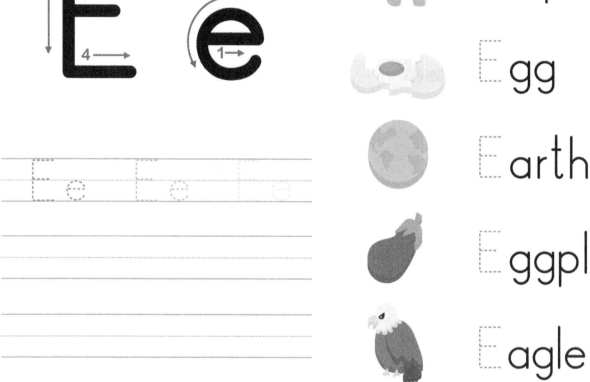

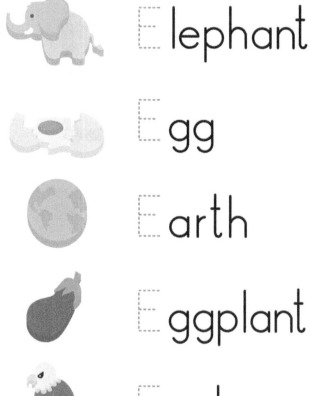

E lephant

E gg

E arth

E ggplant

E agle

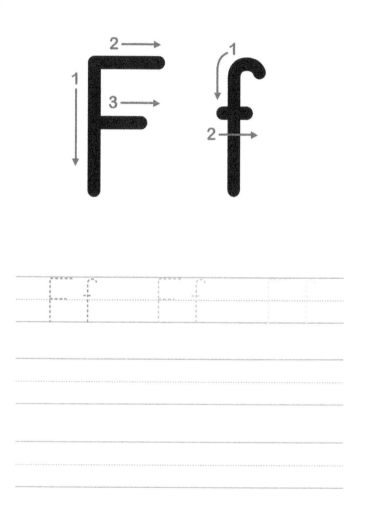

F f

 Fish

 Frog

 Flower

 Flag

 Fork

G g

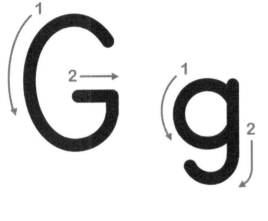

Gg Gg Gg

 G rapes

Gumball

Ghost

Giraffe

Gift

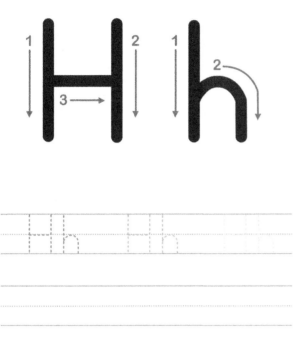

Hh Hh

Hat

Hand

Heart

House

Hippopotamus

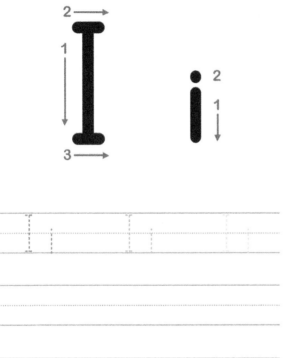

I i

 Ice Cream

 gloo

 guana

 ce

 nk

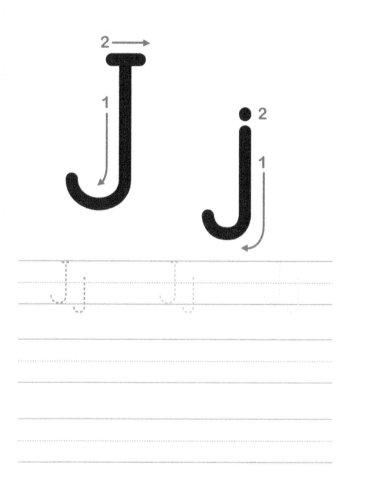

 Jellyfish

 Jam

 Jelly

 Jacket

 Jigsaw

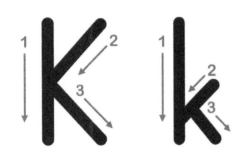

K k

K k   K k   K k

 Kangaroo

 Koala

 Key

 Knife

 Kite

L l

1 ↓ 2 → 1 ↓

 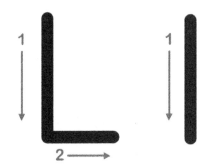L ion

L etter

L emon

L eaf

L adybug

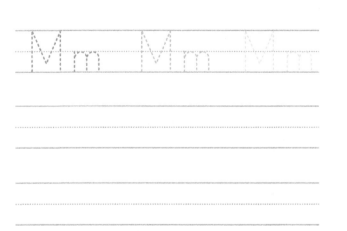

M m   M m   M m

 Moon

 Monkey

 Milk

 Meat

 Mushroom

N n

N n   N n   N n

 Necklace

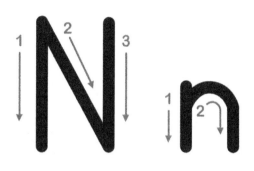 Nest

 Note

 Nail

 Notebook

O o

1   1

○ wl

○ ctopus

○ range

○ strich

○ rchid

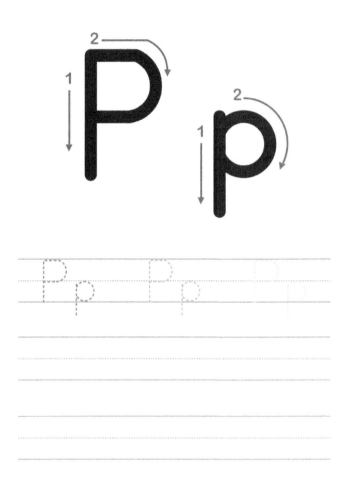

 P an

 P ig

 P encil

 P enguin

 P eacock

Q q

 Question

 Queen

 Quail

 Quilt

 Quill

R r

R r   R r   R r

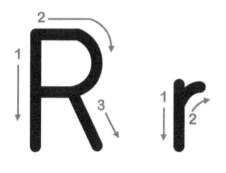

 Rabbit

 Robot

 Rainbow

 Ring

 Rocking horse

S s

S s   S s   S s

 Strawberry

 Shirt

 Slide

 Snail

 Sun

T t

2 →
1 ↓

2 →
1 ↓

 Tiger

 Tablet

 Turtle

 Tennis

 Tree

U u

U
u

 Umbrella

 Unlock

 Unicorn

 UFO

 Underwear

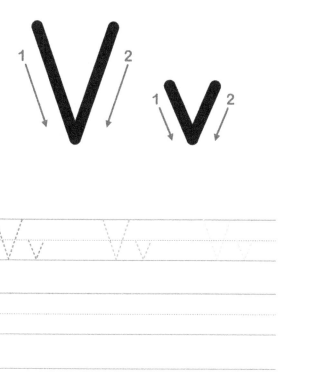

 Vegetable

 Van

 Violin

 Vase

 Volcano

W

1 2 3 4

w

1 2 3 4

Ww Ww Ww

 Whale

 Watermelon

 Wagon

 Worm

 Weathercock

X x

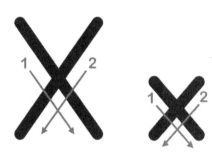

X x X x X x X x

 X-ray

 X'mas **tree**

 Xylophone

 Xerus

 Xiphias

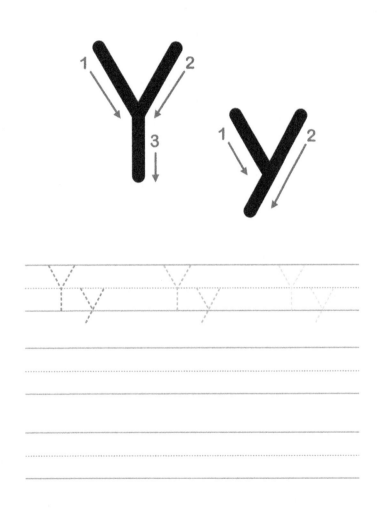

 Y o-yo

 Y ellow

 Y arn

 Y acht

 Y ak

Z z

1 → 2 ↘ 3 →

 Z ip

 Z ebra

 Z eppelin

 Z ero

 Z ucchini

Math

SIMPLE ADDITION

$$\begin{array}{r} 7 \\ + \ 4 \\ \hline \end{array} \qquad \begin{array}{r} 2 \\ + \ 9 \\ \hline \end{array} \qquad \begin{array}{r} 5 \\ + \ 8 \\ \hline \end{array}$$

$$\begin{array}{r} 1 \\ + \ 8 \\ \hline \end{array} \qquad \begin{array}{r} 7 \\ + \ 7 \\ \hline \end{array} \qquad \begin{array}{r} 6 \\ + \ 9 \\ \hline \end{array}$$

SIMPLE ADDITION

$$\begin{array}{r} 9 \\ +\ 5 \\ \hline \end{array} \qquad \begin{array}{r} 9 \\ +\ 3 \\ \hline \end{array} \qquad \begin{array}{r} 8 \\ +\ 5 \\ \hline \end{array}$$

$$\begin{array}{r} 2 \\ +\ 8 \\ \hline \end{array} \qquad \begin{array}{r} 6 \\ +\ 0 \\ \hline \end{array} \qquad \begin{array}{r} 6 \\ +\ 7 \\ \hline \end{array}$$

SIMPLE ADDITION

$$
\begin{array}{r}
7 \\
+\ 5 \\
\hline
\end{array}
\qquad
\begin{array}{r}
8 \\
+\ 6 \\
\hline
\end{array}
\qquad
\begin{array}{r}
4 \\
+\ 5 \\
\hline
\end{array}
$$

$$
\begin{array}{r}
4 \\
+\ 1 \\
\hline
\end{array}
\qquad
\begin{array}{r}
2 \\
+\ 3 \\
\hline
\end{array}
\qquad
\begin{array}{r}
4 \\
+\ 2 \\
\hline
\end{array}
$$

SIMPLE ADDITION

$$\begin{array}{r} 4 \\ + 1 \\ \hline \end{array}$$
$$\begin{array}{r} 2 \\ + 4 \\ \hline \end{array}$$
$$\begin{array}{r} 6 \\ + 3 \\ \hline \end{array}$$

$$\begin{array}{r} 5 \\ + 8 \\ \hline \end{array}$$
$$\begin{array}{r} 6 \\ + 2 \\ \hline \end{array}$$
$$\begin{array}{r} 5 \\ + 1 \\ \hline \end{array}$$

Name: _____   **Score:** _____

SIMPLE ADDITION

$$\begin{array}{r} 8 \\ +\ 4 \\ \hline \end{array} \qquad \begin{array}{r} 7 \\ +\ 3 \\ \hline \end{array} \qquad \begin{array}{r} 6 \\ +\ 2 \\ \hline \end{array}$$

$$\begin{array}{r} 2 \\ +\ 6 \\ \hline \end{array} \qquad \begin{array}{r} 3 \\ +\ 7 \\ \hline \end{array} \qquad \begin{array}{r} 4 \\ +\ 1 \\ \hline \end{array}$$

SIMPLE SUBTRACTION

$$
\begin{array}{r} 3 \\ -\ 2 \\ \hline \end{array}
\qquad
\begin{array}{r} 5 \\ -\ 1 \\ \hline \end{array}
\qquad
\begin{array}{r} 8 \\ -\ 0 \\ \hline \end{array}
$$

$$
\begin{array}{r} 4 \\ -\ 1 \\ \hline \end{array}
\qquad
\begin{array}{r} 5 \\ -\ 2 \\ \hline \end{array}
\qquad
\begin{array}{r} 6 \\ -\ 3 \\ \hline \end{array}
$$

Name: **Score:**

SIMPLE SUBTRACTION

$$
\begin{array}{r} 9 \\ -\ 2 \\ \hline \end{array}
\qquad
\begin{array}{r} 10 \\ -\ 1 \\ \hline \end{array}
\qquad
\begin{array}{r} 5 \\ -\ 4 \\ \hline \end{array}
$$

$$
\begin{array}{r} 8 \\ -\ 2 \\ \hline \end{array}
\qquad
\begin{array}{r} 9 \\ -\ 3 \\ \hline \end{array}
\qquad
\begin{array}{r} 7 \\ -\ 4 \\ \hline \end{array}
$$

Name: **Score:**

SIMPLE SUBTRACTION

$$
\begin{array}{r} 11 \\ -\ 2 \\ \hline \end{array}
\qquad
\begin{array}{r} 8 \\ -\ 4 \\ \hline \end{array}
\qquad
\begin{array}{r} 5 \\ -\ 5 \\ \hline \end{array}
$$

$$
\begin{array}{r} 7 \\ -\ 5 \\ \hline \end{array}
\qquad
\begin{array}{r} 8 \\ -\ 5 \\ \hline \end{array}
\qquad
\begin{array}{r} 12 \\ -\ 3 \\ \hline \end{array}
$$

SIMPLE SUBTRACTION

$$
\begin{array}{r} 7 \\ -\ 7 \\ \hline \end{array}
\qquad
\begin{array}{r} 8 \\ -\ 7 \\ \hline \end{array}
\qquad
\begin{array}{r} 15 \\ -\ 6 \\ \hline \end{array}
$$

$$
\begin{array}{r} 12 \\ -\ 4 \\ \hline \end{array}
\qquad
\begin{array}{r} 8 \\ -\ 1 \\ \hline \end{array}
\qquad
\begin{array}{r} 7 \\ -\ 2 \\ \hline \end{array}
$$

SIMPLE SUBTRACTION

$$\begin{array}{r} 7 \\ -\ 4 \\ \hline \end{array} \qquad \begin{array}{r} 10 \\ -\ 1 \\ \hline \end{array} \qquad \begin{array}{r} 9 \\ -\ 3 \\ \hline \end{array}$$

$$\begin{array}{r} 13 \\ -\ 4 \\ \hline \end{array} \qquad \begin{array}{r} 6 \\ -\ 3 \\ \hline \end{array} \qquad \begin{array}{r} 4 \\ -\ 1 \\ \hline \end{array}$$

COUNT FORWARDS

by 3S

1. 3 , 6 , 9 , 12 , _____ , _____ , _____

2. 1 , 4 , 7 , 10 , _____ , _____ , _____

3. 5 , 8 , 11 , 14 , _____ , _____ , _____ ,

4. 11 , 14 , 17 , 20 , _____ , _____ , _____ ,

COUNT FORWARDS
by 4S

1. 4 , 8 , 12 , 16 , ____ , ____ , ____

2. 2 , 6, 10 , 14 , ____ , ____ , ____

3. 5 , 9 , 13 , 17 , ____ , ____ , ____ ,

4. 7 , 11 , 15 , 19 , ____ , ____ , ____ ,

COUNT FORWARDS

by 5S

1. 5 , 10 , 15 , 20 , ____ , ____ , ____

2. 7 , 12 , 17 , 22 , ____ , ____ , ____

3. 9 , 14 , 19 , 24 , ____ , ____ , ____ ,

4. 8 , 13 , 18 , 23 , ____ , ____ , ____ ,

COMPARING NUMBERS

Write the "<", ">" or "=" symbol

1. 1   3

2. 9   6

3. 4   4

4. 10   3

5. 8   8

6. 5   6

7. 2   8

8. 6   6

COMPARING NUMBERS

Write the "<", ">" or "=" symbol

1. 4   9   **5.** 7   2

2. 2   1   **6.** 8   3

3. 4   7   **7.** 1   8

4. 9   10   **8.** 5   2

COMPARING NUMBERS

Write the "**<**", "**>**" or "**=**" symbol

1.   12   14   **5.**   4   1

2.   2   8   **6.**   5   9

3.   6   3   **7.**   11   8

4.   7   7   **8.**   5   6

COMPARING NUMBERS

Write the "**<**", "**>**" or "**=**" symbol

1. 8   3   **5.** 7   1

2. 4   6   **6.** 8   4

3. 8   7   **7.** 9   8

4. 9   9   **8.** 3   2

COMPARING NUMBERS

Write the "<", ">" or "=" symbol

1. 4   4   **5.** 3   2

2. 2   6   **6.** 12   3

3. 3   3   **7.** 14   8

4. 9   5   **8.** 10   10

COMPLETE THE NUMBERS

1		3	4	5
6	7		9	
11	12		14	15
	17	18		20
21			24	25

COMPLETE THE NUMBERS

6		8	9	10
11	12	13		15
16		18	19	
21		23		25
		28	29	

COMPLETE THE NUMBERS

9		11		13
14	15		17	18
	20	21		23
24		26		28
29		31	32	

Name: Score:

COMPLETE THE NUMBERS

12	13		15	16
17		19	20	
22	23		25	
	28		30	31
	33	34		36

Name: _____   **Score:** _____

COMPLETE THE NUMBERS

16		18	19	
21	22		24	
	27		29	30
	32	33		
36	37		39	40

WORD PROBLEMS

1. Salve picked 2 apples and Tom picked 5 apples. John picked 2 apples. How many apples were picked in all?

2. For Christmas, Faye had 10 pieces of candy. She ate 2 pieces. How many pieces of candy were left?

WORD PROBLEMS

1. Daryl's high school team played 6 basketball games this year. He attended 2 games. How many basketball games did Daryl miss?

2. A book store had 10 books in the bargain bin. If they sold 3 books, but then put 2 more books in the bin. How many books would be in the bin?

WORD PROBLEMS

1. Tom ran for 6 minutes and Ivan ran for 4 minutes. How many minutes did the boys run in all?

2. Kyle had 2 books. If he sold 1 of them and used the money he earned to buy 2 new books, how many books would Kyle have?

WORD PROBLEMS

1. Sheila had 7 playing cards, and 3 were torn. Keith bought 2 of his cards. How many playing cards Sheila have now?

2. A teacher had 7 worksheets to grade. If she graded 4, how many worksheets would she have to grade?

WORD PROBLEMS

1. Gwen had 8 biscuits, she gave 4 to her sister, how many biscuits are left for her?

2. Jan had 10 balloons. 6 are green and the rest are blue. How many balloons are blue?

WORD PROBLEMS

1. Sheena ate 8 cookies while Charie ate 4 cookies. How many more cookies did Sheena eat than Charie?

2. Kenji got 4 new shirts as birthday gifts. If he already 5, how many shirts does he have now?

WORD PROBLEMS

1. Jane has 3 cats. May has 4 cats. How many cats do both of them have?

2. I built 9 snowmen in the field. If 3 melted, how many is left?

WORD PROBLEMS

1. I have 3 pencils in the can. The can fits 10 pencils, how many more pencils can fit?

2. Grace had 10 sweets but she gave Jenna 4. How many is left?

ANSWERS

p.73	11		9		6
	11		13		3
	13		8	**p80**	9
	9		6		4
	14	**p.77**	12		0
	15		10		2
p.74	14		8		3
	12		8		9
	13		10	**p.81**	0
	10		5		1
	6	**p.78**	1		9
	13		4		8
p.75	12		8		7
	14		3		5
	9		3	**p.82**	3
	5		3		9
	5	**p.79**	7		6
	6		9		9
p.76	5		1		3
	6		6		3

Made in the USA
Coppell, TX
27 August 2021